The
Garden

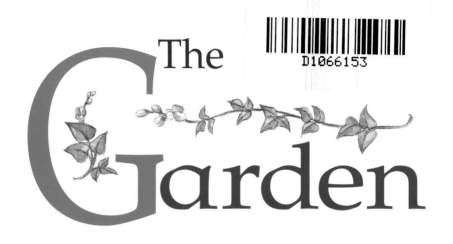

By Veronica Freeman Ellis
Illustrated by Holly Cooper

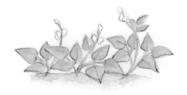

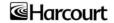

Harcourt

Orlando   Boston   Dallas   Chicago   San Diego

www.harcourtschool.com

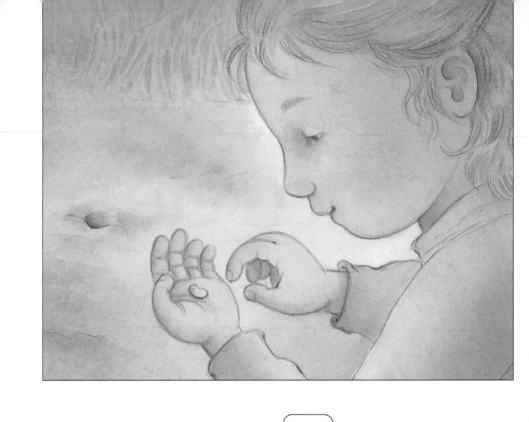

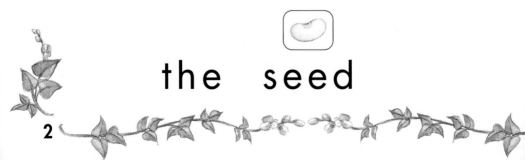

the seed

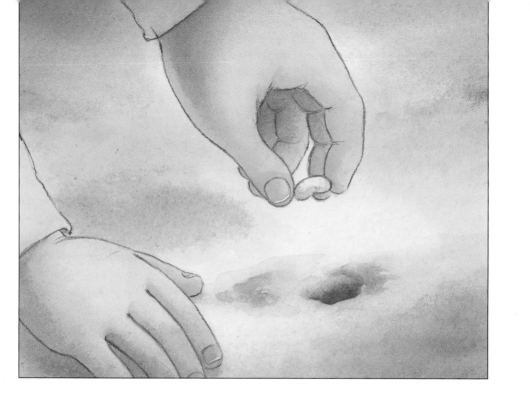

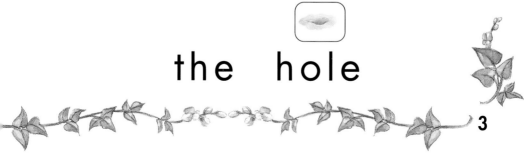
the hole

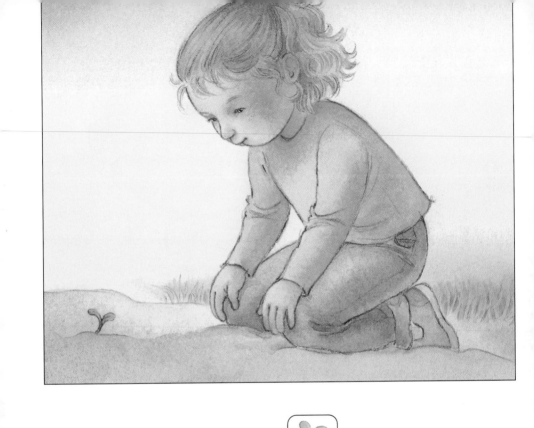

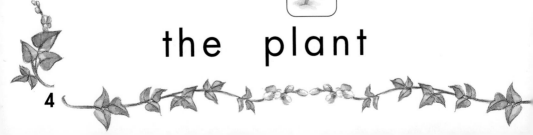

the plant

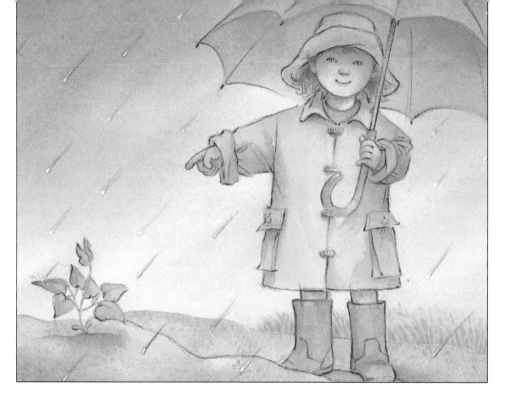

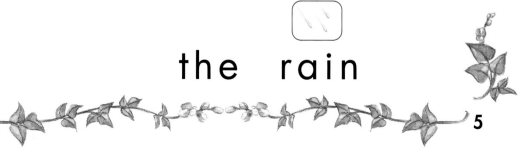

the rain

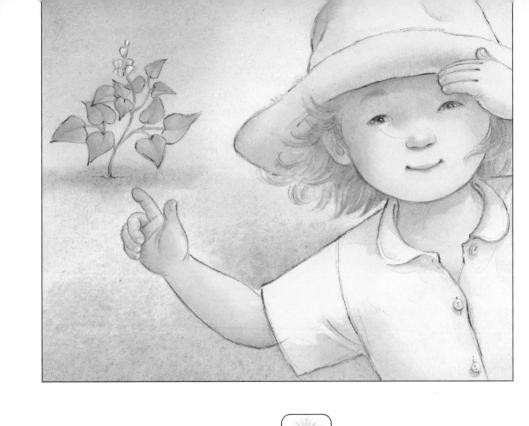

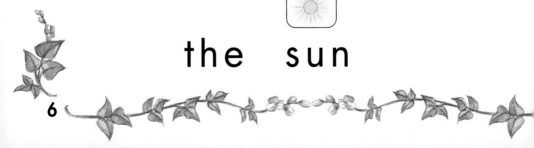

the sun

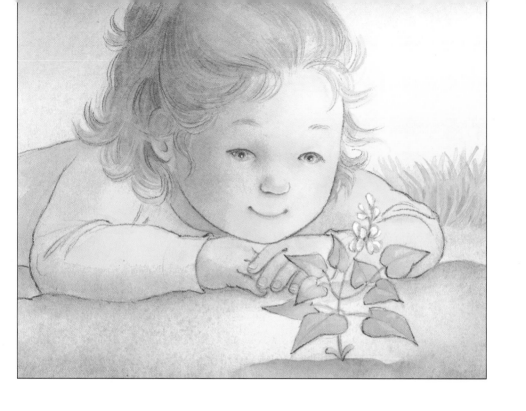

the 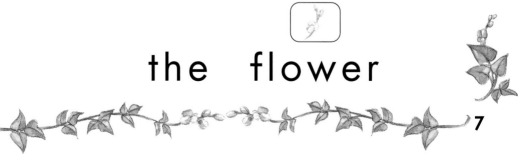 flower

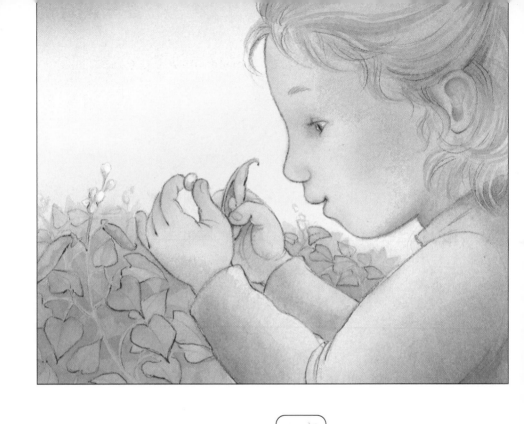

the garden